Starting from Your Head

Mental Number

by David Fielker

ISBN 1 874099 17 0

Typeset and designed by Fran Mosley
Cover design by Simon Spain, Orangebox Editions
Printed by GPS Ltd, Watford

The BEAM Project is supported by
Islington Education

CONTENTS

WHY MENTAL NUMBER?

Mental work in number can be much more than merely calculating in one's head.

Calculating in itself can, when performed mentally, reveal for the teacher a great deal about the children. When performing mental calculations children demonstrate an ability to be inventive, using algorithms they have never been taught. They also show an ability to use the knowledge they have to derive new knowledge. A typical example is the seven-year-old who wanted to calculate 7 x 7. He knew that three sevens were 21. Therefore six sevens, he said, were twice as much — 42. Add another seven — 49. Thus he developed an original algorithm, using ideas about the relationships between multiplication and addition, to get from the known fact, 3 x 7 = 21, to an unknown fact, 7 x 7 = 49.

But there are other virtues in dealing with number in your head. Children build up a different sort of familiarity through their mental images of number. They attain a greater flexibility in all aspects of number work by being able to manipulate those images. They can work towards generalisation of number patterns by visualising the generality — see the section on *Sequences and Formulas*. And it is only in their minds that they can form an idea of infinity.

Children of all ages and abilities have their own images, according to their knowledge and experience, but we as teachers do not always discover what goes on in their heads.

Hence the importance of *discussion* as an accompaniment to mental work. Then not only does the teacher develop insight into what is happening, but the children themselves can develop their own insights further by having to make them explicit, they can share their ideas and their methods and thus make them available to others, they can form judgements about the alternatives, and generally they can develop a greater flexibility and familiarity with number.

An important idea to point out is that there are different forms of stimulus to mental activity. These can be divided into *aural* and *visual*.

4

Aural stimuli are perhaps more common than visual ones in traditional types of mental calculation, with the questions dictated by the teacher and the answers given orally or in writing. This sort of stimulus leaves the children free to manipulate in their minds the images they have of the numbers, and can give rise to completely different methods of working. When the numerals can be seen, it can impose a more conventional algorithm, especially if the numerals are laid out in a formal way, as equations:

$$34 + 27 =$$

or in a stylised layout:

$$\begin{array}{r} 34 \\ +27 \\ \hline \end{array}$$

(One used to talk of 'horizontal' or 'vertical' addition!)

Reliance on *known* facts, like the multiplication tables, often depends on an aural input: you have to hear the sounds in order to give an automatic response. To realise this, you only have to try responding to questions about multiplication facts in a foreign language: you have to translate into English in order to calculate, then translate back in order to reply. This realisation should make teachers more understanding of the delays experienced from those children whose first language is not English, and who have learnt their number bonds in their first language.

In general in number work, visual stimuli are probably far more common than aural ones, since *any* activity in number, be it working with numerals or with some form of apparatus, involves to some extent some mental work. This mental work can consist of calculations which are subsidiary to written calculations, like the recall of addition facts for single digits when adding two-digit numbers. It may involve the mental manipulation of objects or structural apparatus. It may concern the visualisation of general cases when deriving a formula. Or it may be some use of those personal mind pictures that all children have.

Organisation

Some short sessions can be planned specifically for mental number work, either with a whole class or with groups. As discussion is an important feature, whole class sessions need to be handled with care, both to gauge the concentration span (which may depend largely on the age of the children) and to ensure that all discussion is audible.

One technique to overcome the problems of the shy or quiet speaker is for the teacher to repeat what is said, but in doing so she or he must be careful not to let a rephrasing, an emphasis or a tone of voice add to or alter what has been said.

Children will need to explain their methods, and access to a flipchart, blackboard or overhead projector will help. The recording may be done by child or teacher, but if the latter then the teacher must be very careful not to impose a structure by the way she or he writes things down. If, for example, two numbers are to be added, then (as indicated above) whether they are written underneath each other or in the same line with a plus sign between them can make a crucial difference to the method used.

The choice between whole class or group work is always difficult. Whole class work means that everyone has access to the ideas that are being discussed, but it can be dominated by the more vociferous, and the more passive can lose interest. If the children are all working in groups, this means that more children can participate, but the teacher then loses control of most of the situations, as well as the chance to hear most of what is being said, and therefore some of the opportunities for assessment. A useful compromise sometimes is to pose a problem to the class, have it discussed in groups, and then ask groups briefly to report back to the class. But because of the memory span involved this works better with older children.

Apart from considerations of maturity, children may not be used to discussion of methods. In this case a little at a time in the early stages (perhaps individually or in small groups) will help them gradually develop the skills of explaining and listening. Obviously these are skills that will be developed right across the curriculum, in other subjects as well as in mathematics. But mathematics, and number work in particular, has its own traditions that have often militated against freedom of choice, explanation of methods and evaluation of the alternatives. Sometimes these traditions will gradually have to be broken down.

THE
ACTIVITIES

IMAGES

Children's images are important because they affect their feelings of familiarity with numbers. The images also affect the ways in which children calculate mentally.

There seems to be a great difference in how anyone imagines something as simple as the number five. The main difference is between those who see the numeral, and those who see the number in some concrete way: the pattern on a dice, five objects, a yellow rod, a position on a number line or in a 100-square.

With larger numbers there may be a tendency to see them in more abstract ways. But much depends on two things: whether one is being asked to calculate with the numbers, and what one's experience is.

Before children know number facts automatically, calculation with small numbers may rely on the concrete images, with visualisation in terms of whatever apparatus the children find familiar. (Sometimes they turn the numerals themselves into concrete images; as the girl says in the Peanuts cartoon, when faced with '7 + 3':

"Anything with a 3 is easy because you just take the first number and then count the little pointy things on the 3".)

Calculation, even with two-digit numbers, often relies on the visualisation of a written algorithm, and children who work this way generally have an understanding of the notation, and are consequently able to manipulate that, instead of relying on their former concrete images. But, as the next section on *Methods* will show, children have surprising ways of avoiding *standard* algorithms

"What do you 'see' when you hear the word 'five'?"

when they calculate in their heads, and this again generally shows an understanding of the notation.

Children's experience with structural materials may influence their mental images of larger numbers to some extent, but the size of the numbers involved may outstrip the experience, so to speak. A thousand is easier to picture in terms of multibase blocks than in terms of Cuisenaire rods. But a million may be beyond anyone's actual experience, and may have to be *constructed* mentally. The teacher can help by asking, if appropriate, how many 1000-blocks are needed, and how they could be arranged; or asking about the number of millimetre squares in a square metre and encouraging the visualisation of those in sections.

Even so, it is not necessary to have a concrete image of a number, and it will also be useful, both from the point of view of familiarity and from the point of view of calculation, to be able to conjure up the *numerals* and work with those. The whole idea of place value involves working with numerals rather than with objects, and the numerals themselves eventually become the mental images on which one can operate.

The activities here which deal with the mental manipulation of a small number of objects mirror those which can be done practically with counters or other objects. They are part of an alternative way of becoming familiar with number facts concerning addition and subtraction, and multiplication and division. Instead of giving two numbers and asking for the result of an operation performed on them, one number is given, to be *analysed* in terms of how it can be made up. For instance, it is certainly a worthwhile activity for children to take, say, 15 actual counters and try to arrange them in rows of 2, 3, 4, 5, and so on. Doing this mentally, however, not only encourages children to sharpen up their powers of imagery with numbers and with objects; it also gets them to work towards combining this with some mental calculation. Maybe they can 'see' 15 counters arranged in three rows; maybe they know that three fives

IMAGES

are 15, and *then* see the three rows of 5. Whether the children do this practically or mentally, the principle of analysing a number in this way is an important one. It encourages far more flexibility with numbers than does the more usual technique of presenting 'sums'.

Note that part of this flexibility is to include, say, 15 put into rows of 2 or of 4, with consequent 'remainders'. It is a false economy to introduce ideas about division by restricting it to cases where it works out exactly and to defer the discussion of remainders until later.

Note also the difference between 'arrange 16 in two rows' and 'arrange 16 in rows of two'. Each of these situations will sooner or later be described as '16 divided by 2', but the images are quite different. They correspond to the two aspects of division (known technically as 'partition' and 'quotition') which are typified respectively by the questions 'Share 16 between two,' and 'How many twos in 16?'

It is also appropriate to include some prime numbers (which can only be put in one row or rows of one, so to speak) and eventually to identify these.

The first set of activities should not be dealt with all at once, and obviously the size of the numbers should suit the ability of the children. Imagining 5 may be a short, one-off activity with a group or a class, and the images can be shared, both so that the teacher knows what is going on, and so that the children can hear about other images that they may not have considered. Maybe it is appropriate then to consider, say, six counters, so that differences between children's images of the two numbers can be compared. For older children it may be appropriate to consider their images of, say, 1000, 10000, and so on up to a million, or to consider such numbers as 12,345.

The activities which put counters in rows can be combined with practical work with counters, as a precursor to it, as a follow-up to it, or during the practical activity when the teacher occasionally can interrupt and say, "Close your eyes and *imagine* what you are doing".

1000

Close your eyes

What do you 'see' when you hear each of the following?

5, 10, 12, 1, 24, 100,

a thousand, a million,

a half, a third,

36, 56, 23, 365

Imagine five counters

Arrange them in a domino pattern.

Arrange them in two groups in different ways.

In three groups.

Arrange them in other ways. What does each way tell you about 5?

Do this with other numbers of counters.

Imagine six circles

Arrange them in two rows.

In three rows.

In rows of two.

In rows of three.

Imagine sixteen circles

Arrange them in rows of two.

In rows of four.

In rows of eight.

In rows of three.

Imagine other numbers of circles

and do the same thing.

METHODS

Two of the things said earlier need to be repeated here:

i   children calculating in their heads use a far greater variety of methods than they do when they are working on paper;

ii they tend to use different methods according to whether the questions are aural or written.

What is also evident is that in order to invent alternative algorithms, children rely on their familiarity with numbers, a knowledge of place value, and a knowledge of the properties of the four operations. These abilities do not always reveal themselves in more formal testing procedures, and so the explanation of methods is also a valuable way in which assessment may take place.

It is also true to say that children's ability to invent and use different methods is dependent on this sort of knowledge. This has two consequences. One is that the

One child's visualisation of '4 + 3'

teacher can assess the children's own methods in the light of the knowledge they can be expected to have. The other is that the teacher can be aware of what sort of knowledge is going to be useful in the construction of algorithms, and so set out to facilitate the acquisition of it.

As has already been suggested, in any situation in which mental calculations are being done, formal or informal, it is invaluable for the children to explain what they have done. The process of explanation, besides being a useful skill in itself, often makes the child doing the explaining more explicitly aware of the method he or she has used. It is also an important skill to be able to understand the explanations of others.

It is also a valuable exercise to consider the advantages and disadvantages of different methods in different circumstances. One should at the same time bear in mind that any method is not only peculiar to the particular question, but it may also be peculiar to the person doing the calculating. The object is not necessarily to find a best method for anything that everyone should adopt, but to share experiences, methods and images, and thus broaden the options for everyone and widen the range of insights available.

The questions suggested here are only a selection of possibilities, but they indicate the sorts of questions that produce different strategies.

METHODS

Do these calculations in your head

and explain how you did each one:

3 + 4	13 + 4	13 + 14
23 + 54		

8 + 7	9 + 4	5 + 6
15 + 15	15 + 17	34 + 57

24 + 26	99 + 101	39 + 42

10 − 3	11 − 3	21 − 9
21 − 19	100 − 47	110 − 103

53 − 27	48 − 24	30 − 3
1000 − 1		

3 x 7	6 x 7	8 x 4
4 x 8	7 x 7	12 x 4
12 x 7	13 x 6	13 x 8
12 x 12	13 x 14	

Do these calculations in your head

and explain how you did each one:

36 ÷ 3	42 ÷ 3	100 ÷ 25
100 ÷ 5	500 ÷ 25	246 ÷ 2
338 ÷ 2	336 ÷ 4	105 ÷ 5
205 ÷ 5	215 ÷ 5	432 ÷ 4
432 ÷ 3		

Did you do similar calculations the same way?

Could you have done each in a different way?

Are some ways better than others? When? Why?

I keep them in my head then I close my
eyes and see whats in my head whats
happening I probably see hands I count on 19.

Tayo

the way I worked it out was there is a ten in 19 so all I had
to do was add the ten to the 34 and that would give you
44 so all you had to do was add the 9 and the 4 together and
that would give you 43 53.

Toyos

$$30 \atop \bullet \, {4 \atop 0} + {15 \atop 0} \, {4 \atop \bullet} = 53$$

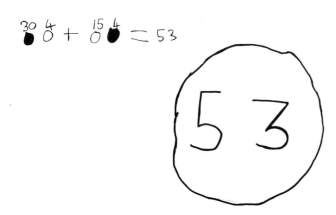

53

Four children show
how they worked
out '34 + 19'

34 + 19
30 + 4 + 19 = 53

I would work it out like this
I would get three 10's then + 4
then + 15 which makes 48
then + another 5.   Vian

COUNTING

There are two sorts of questions here.

The first group is about counting people, but in situations where the people are not necessarily arranged in some convenient order, and cannot physically be put into such an order; so an order has *mentally* to be imposed. There are various ways of doing this in a classroom: perhaps counting a table at a time, or doing a 'sweep' of the room. Children will have their own methods of organising the count. The hall and the playground may pose other problems, and hence other solutions.

The problem of counting people is quite different from the problem of counting similar numbers of objects, because objects can generally be moved around and organised into physical groups. However, if the 'objects' are pictures of objects pasted onto a card, or printed on the page of a book, then the problem is similar to that of counting people in a room, in that the organisation has to be done mentally.

The rest of the questions involve counting by calculating, so to speak. Straight counting from 1 upwards is a familiar exercise, but to start somewhere else is not always so easy for younger children, especially when the count crosses an awkward boundary like 100. And counting in twos, threes, fives or tens is often a useful technique anyway because it is quicker.

However, counting in, say, threes, starting from some number other than zero is an exercise in calculation — that is, in successive addition — rather than in using the multiplication tables. Some interesting patterns will be produced, in sound as well as in mind pictures. For instance, counting in fives from 2 produces

2, 7, 12, 17, 22, 27, 32, 37, . . .

and counting in nines from 10 produces

10, 19, 28, 37, 46, 55, 64, 73, . . .

Being able to count numbers is something not often required. The raffle ticket problem is however a very practical one!

How many people are in the classroom?

How many people are in the hall?
How many people are in the playground?
How do you count them?

Count:

in ones, from 89 to 103;
in twos, from 2 to 50;
in twos, from 1 to 31;
in threes, from 3 to 51;
in threes, from 2 to 50;
in threes, from 1 to 49;
in fives from 5; in fives from 2;
in tens from 10; in tens from 3.
in nines from 9; in nines from 10;
in eights from 2; in eights from 5.

Count backwards:

from 20; from 50;
in twos from 50; in twos from 51;
in threes from 36; in threes from 40;
in fives from 50; in fives from 57;
in tens from 100; in tens from 97;
in nines from 100.

How many whole numbers are there from 1 to 10?

From 10 to 100?
From 100 to 1000?
From 57 to 84?

A book of raffle tickets is numbered from 500 to 599

Some have been sold, in order, and the next ticket available is numbered 543. How many have been sold?
How many are left?

How many different whole numbers can appear on the eight-digit display of a calculator?

What if the 9 key did not work?

EQUIVALENCES

This section is intended as an encouragement to children to make explicit some of the mental techniques that they already use.

Thus they will often replace 99 + 46 by 100 + 45, although they may not think of it this way. To encourage more flexibility, we do not require the answer here; we merely want to be able to say that

99 + 46 = 98 + 47 = 97 + 48 = 96 + 49 = . . .

or

99 + 46 = 100 + 45 =101 + 44 =102 + 43 = . . .

It so happens that 100 + 45 is probably the easiest of these to calculate directly.

What is also important is how each particular operation affects the construction of the equivalences. If the operation is addition, the increase in one of the numbers is compensated for by a decrease in the other.

With subtraction it is different: both numbers must increase or decrease together. Now the choices can vary more. 134 − 98 may best be replaced by 136 − 100, and 567 − 101 may be easier as

566 − 100, but perhaps 1000 − 3 is most convenient as 999 − 2. There is a lot of opportunity for discussion! There is also a connection with the rounding off that is discussed in the section on *Estimation* (p.22).

Multiplication and division have similar characteristics. Thus 24 x 25 can be replaced by

12 x 50,   6 x 100,   3 x 200,   120 x 5;

and 210 ÷ 42 can be replaced by

105 ÷ 21,   70 ÷ 14,   35 ÷ 7,   30 ÷ 6,

15 ÷ 3,   10 ÷ 2

as well as by many others.

The techniques for calculation thus rely not just on the children's familiarity with the numbers, but also on their knowledge of the nature of each of the operations.

Again, the questions given here are just a selection.

For each of these, what other additions give the same answer?

99 + 46	48 + 72
999 + 478	498 + 234
538 + 345	

For each of these, what other multiplications give the same answer?

4 x 84	3 x 48
2 x 33	8 x 113
256 x 121	35 x 41

For each of these, what other subtractions give the same answer?

134 − 98	567 − 101
1000 − 3	1908 − 1902
3794 − 1749	

For each of these, what other divisions give the same answer?

34 ÷ 4	250 ÷ 50
248 ÷ 14	345 ÷ 9
455 ÷ 35	783 ÷ 81

$$1000 - 3$$
$$1001 - 4$$
$$1002 - 5$$
$$2002 - 1005$$
$$3002 - 2005$$

Part of one child's work on equivalences

CALCULATIONS

It is possible merely to bombard children with a set of miscellaneous calculations to be performed mentally.

The questions here are different in that they are structured in some way, and this gives them more point and interest.

The structure may also give rise to some other characteristics that may stimulate some investigation. For instance, starting with 1 and adding successive odd numbers produces the square numbers.

A third advantage is the 'and so on' that ends each group. Here is an implicit invitation to the children to carry on as far as they can, a challenge! It is surprising how well children will rise to this bait. I have known some pupils continue doubling numbers way beyond what I would have expected.

The subtractions may stop naturally when there appears to be not enough left to subtract from, but we know from similar work with calculators that many children are quite capable of developing the idea of negative numbers in this sort of situation.

A young child's image for '4 add 2'

20

Start with 1

Double it.
Double again. Keep doubling. How far can you go?

Start with 1

and keep trebling.

Start with 1

Add 2, add 3, add 4, and so on.
How far can you go?

Start with 1

Add 3, add 5, add 7, and so on.

Start with 1

Add 3, add 4, add 3, add 4, and so on.

Start with 1

Add 7, subtract 3, add 7, subtract 3, and so on.

Start with 100

Subtract 1, then 2, then 3, then 4, and so on.

Start with 100

Subtract 5, then 2, then 5, then 2, and so on.

Start with 1

Halve it. Halve it again. And so on.

Try the same thing with thirds

half

ESTIMATION

The first set of extracts from a lesson with a class of 10-year-olds is taken from the author's *Using Calculators with Upper Juniors* (Association of Teachers of Mathematics, 1985). The children had found earlier that entering a number on a calculator and then pressing ☒ ⚌ produced the square of the number. The teacher had presented them with various numbers, and asked what needed to be squared to get them. These problems they had solved by trial and error, using their own calculators.

The 10 billionth hamburger — see page 28

DF I thought we'd do it differently this morning.

All Without calculators?!

DF Yes, but I'll help you, because I'll do the working out. Now, I squared something, and got that.

(*He writes*

15129

on the blackboard.)

It's all right. You tell me what number you want to try, and I'll work it out for you. But you'll have to agree about which number you try.

This can be a valuable technique when using calculators. When children have their own (which they should do frequently) there are two disadvantages: there tends to be little discussion, and the teacher has no idea what is going on! This way discussion must take place, and the teacher can hear it and react to it.

John 71.

There is general agreement to this. The teacher squares 71 and writes:

71   5041

Gary   201.

Cheryl   182. No, that's even.

Jonathan   233, because 5041 is about one third of the number on top (*15129*) and I timesed 71 by 3.

The teacher writes:

233   54289

Several   139 . . . 232 . . . 137.

Paula   217.

Shelley   I think that's too much.

John   133, because it's an odd number.

Joshua   93, because it gives us a small number.

Jonathan   3 goes into 15129, so we're looking for a number that 3 goes into.

Joshua   Like 93.

DF   What sort of number do we want?

Cheryl   Somewhere around 200.

Jonathan   It must be over 100, because a hundred hundreds is ten thousand, and that's less than that number.

A voice   142.

Cheryl   It's more than 142, because 71 and 71 comes to 142, and that number (*5041*) added together (*she means doubled*) is less than 15129.

Jonathan   You're squaring it. You're not timesing it.

Jonathan has now rejected his own earlier hypothesis of proportionality (the number must be 3 times 71, because 15129 is about 3 times 5041) and contradicts Cheryl's similar idea.

There is now a discussion about whether 141 or 142 is a multiple of 3, beginning with a comment that 3 goes into 21.

Cheryl   21 and 21 is 42.

DF   So 3 goes into 42, does it?

Cheryl   It doesn't go into 41.

ESTIMATION

DF Does that mean 3 goes into 142?

Jonathan No. 3 doesn't go into 100, but it goes into 99, and if you add 42 onto that it leaves you 141.

DF All right, is it? Worth trying 141?

All Yes.

The teacher writes

141   19881

Cheryl It's got to be underneath that number.

A voice 133.

Jonathan 3 doesn't go into it.

Several It does! It doesn't!

Jonathan 3 doesn't go into 100, so it doesn't go into 133.

Cheryl It'll go into 132, then.

Much is going on, and it is all being done mentally. Apart from various ideas concerning recognition of multiples of 3, note the estimation that takes place: Jonathan's "about one third", Shelley's "I think that's too much", Cheryl's

"somewhere around 200" and Jonathan's rejoinder, "It must be over 100, because a hundred hundreds is ten thousand".

*   *   *   *   *

The next extract illustrates a different idea.

Amanda was working from a textbook, and was using a calculator to answer the question: *How long will it take to pay for a carpet costing £46.70 if £5.70 is paid at the time of the purchase and the rest is paid at 50p a week.*

She had calculated $41 \div 50 = 0.82$, so she had done the subtraction and correctly divided. But she was having trouble interpreting the 0.82 on the display.

Amanda Is it 82 weeks? It seems too long.

DF (*after a pause*) What would happen if you paid £1 a week?

Amanda (*after a little thought*) 41 weeks. Oh yeah, it's 82 weeks.

These anecdotes illustrate two aspects of estimation.

If you are dealing with abstract numbers, with no external context, then the only way to estimate is to do a 'rough' calculation. Jonathan knew that 5041 was about one third of 15129 presumably because 500 was one third of 15000 (or perhaps even because 5040 was one third of 15120). And because 100 squared was 10000, he knew that the square root of 15129 had to be more than 100.

On the other hand, Amanda had some intuition about 82 weeks being too long for the repayments. It sounded unreasonable to her in the *context*. Contextual estimation generally depends on a knowledge of the real world, and on some aspects of measurement. There is something wrong if you have worked out that a cyclist is pedalling at 190 kilometres per hour, or a bar of chocolate costs £253, or the school is a kilometre high. You know that cyclists travel more slowly than cars, that chocolate bars cost around 25p, and that the three-storey school with rooms just over three metres high must be about ten metres high.

However, we often wish to be a little more precise about our estimating, and then we resort to some sort of *approximation*.

We usually do this by *rounding off* the figures involved. In other words, we can replace each figure by something close, but which is easier to deal with.

If we cannot remember whether eight nines are 63 or 72, then we can say that eight tens are 80, so 72 is more likely. We may compensate for the nine being raised by lowering the eight to seven, and seven times ten is 70, which makes us feel even better about the 72.

Children can gradually learn how adjustments like this can become increasingly sophisticated, and also that they depend on which operation is involved. An approximation to 123 x 417 will be 100 x 400 = 40,000, but this will be an *underestimate*. An approximation to 179 x 487 will be 200 x 500 = 100,000, but this will be an *overestimate*. An approximation to 187 x 216 of 200 x 200 will be closer, but we cannot be sure whether it

ESTIMATION

will be too big or too small.

On the other hand, an approximation to 5719 ÷ 213 of 6000 ÷ 200 = 30 will be an *overestimate*, because we are dividing more by less.

If we have multiplied 237 by 594 on a calculator and made it 22278, and we are checking mentally whether our answer is reasonable, then we probably multiply 200 by 600. This gives 120,000, and it looks as if we are a long way out! We can try the calculation again, and get 140778, which looks much better. In fact 22278 is the result of multiplying 237 by 94, so perhaps the 5 from the 594 did not register on the calculator. It helps to press the buttons firmly, and check that the correct number has registered in the display before pressing the next key. (Note also that it helps to know why you went wrong, rather than just making the correction, so that the error can be avoided in future.)

Since it is fairly easy to press the wrong keys on a calculator, it has always been recommended that calculations performed on one should be checked in some way. Strangely, that recommendation has never been made with the same vehemence about calculations on paper or in the head, where other things are far more likely to go wrong!

Estimating, approximating and checking answers should be a normal part of solving problems. Generally it helps to develop the right attitude if children's answers are met, not with a pronouncement from the teacher that they are right or wrong, but with queries about what the children think. "Do you think it's right?" "Is it reasonable?" "How do you know it's right?" "How could you check it?"

It can give children a wrong impression if work on estimation is only carried out on particular occasions. However, there are certain techniques and ideas that will profit from an occasional concentrated discussion. The activities suggested below come into this category.

Each of the following calculations can be estimated by the corresponding approximations

Decide whether they are underestimates or overestimates, or that you cannot tell.

417 + 378	➡	400 + 400
417 + 305	➡	400 + 300
488 + 378	➡	500 + 400
488 + 305	➡	500 + 300
894 − 209	➡	900 − 200
894 − 285	➡	900 − 300
823 − 209	➡	800 − 200
823 − 285	➡	800 − 300
504 x 625	➡	500 x 600
589 x 625	➡	600 x 600
589 x 693	➡	600 x 700
504 x 693	➡	500 x 700

3275 ÷ 62	➡	3000 ÷ 60
3875 ÷ 62	➡	4000 ÷ 60
3275 ÷ 68	➡	3000 ÷ 70
3875 ÷ 68	➡	4000 ÷ 70

A 500g packet of corn flakes costs 85p and a 750g packet costs 125p

Which is the best buy? Investigate in your local supermarket.

Discuss the following

"1,250,000 of you took the trouble to fill in our questionnaire."
(British Gas advertisement 4.2.1990)

"Between 1991 and the end of the century the human race is likely to increase by almost one billion (one thousand million) people . . . it could rise to 14 billion."
(*Education Guardian* 10.9.1991)

27

"10 billion hamburgers sold."
(McDonald's advertisement, USA, 1992)

"An estimated £600 million is handed over in pocket money every year."
(*Moneycare* Autumn/Winter 1990)

"London is one of the least traffic-jammed cities in the world. Her cars move along at a sizzling 12.43 miles an hour, while in New York they could only manage 9.36 and in Athens 4.78."
(*Evening Standard* 8.1.1991)

"Thirsty Britons splashed out £126 million on bottled water last year . . . And in the 1990s, experts reckon companies such as Schweppes and Perrier will be tapping us for £200 million a year."
(*Sun* 4.5.1989)

"One in eight teachers changed jobs last year . . . nearly 47,000 teachers resigned their posts."
(*Guardian* 26.9.1991)

Find other examples in the newspapers.

SEQUENCES AND FORMULAS

This section is about algebra, which stems partly from the continuation and observation of number patterns and the subsequent generalisation of them. (For a more detailed discussion, see the article *Algebra* in the BEAM pack, *Spot the Pattern*.)

All the examples here are practical ones, and the mental work can, if one wishes, be left until the generalisation stages: "Close your eyes and imagine more squares being added". Or the whole thing can be done mentally. It depends on the age and experience of the children, and, indeed, on the preference of the teacher!

There are two sorts of generalisations that can be made. One is to determine the rule for how the pattern continues. In the first example, three matches are added each time.

But the crucial generalisation here is to be able to give a rule for the total number of matches, given any number of squares. For 100 squares, there were four matches for the first square, and then we have to add three matches for all subsequent squares: $4 + (99 \times 3)$. One would not expect 100 squares actually to be constructed with matches; it has to be a construction in the mind.

A request for a rule, in words, may produce something like, "You start with 4, then you add 3 times one less than the number of squares you want." After much activity like this, perhaps over a long period, it is eventually a short step to using a letter to represent the number of squares.

Incidentally, the simplifications from

$$4 + (99 \times 3) \quad \text{to} \quad 1 + (100 \times 3)$$

or from

$$4 + 3(n - 1) \quad \text{to} \quad (1 + 3n)$$

need not in the first instance be an 'algebraic' one in the formal sense. If asked how you construct *one square*, children can see that they can start with one match and add three matches. The consequent rule is perhaps easier to deal with. They can then, if they wish, use formal algebraical manipulation to 'prove' that the two rules are equivalent.

SEQUENCES AND FORMULAS

Some of the later examples are more complicated, in the sense that the number patterns tend to take over from the situations from which they came. What is crucial is *how the items are counted*. In the questions about 'frames', for example, it is usual to draw several examples and count the number of squares in the frame each time, making appropriate lists, and then examine the patterns involved and try to make sense of *them*. There are two disadvantages in this. One is that the counting is done in any fashion round the frame, and this can obscure how the situation is structured. The other is that any subsequent hypotheses are based on the number patterns themselves, rather than on the structure of the original situation.

Instead, children can be asked to *imagine* a rectangle and count the squares round it. If it is a 10 by 11 rectangle, they can perhaps register, without counting and *without adding them together*, two rows of 10, two rows of 11, and four squares for the corners. The generalisation is already

inherent in this organisation of the counting, which in a sense is not so much counting as adding the component parts. It is now a small step to a rule for *any* size rectangle: twice the length, twice the breadth, plus 4.

A child's image of numbers in sequence

Imagine four matchsticks making a square

Add three matches to make it into two squares, side by side; how many matches are there altogether?

Add three more to make another square by the side; how many matches now?

Carry on. When you have made 10 squares in a row, how many matches are there?

100 squares? *Any* number of squares?

Imagine three matchsticks making a triangle

Add matchsticks to make another triangle joining it at the side.

Add another triangle at the side. Keep adding triangles, counting the matchsticks as you go.

How many matchsticks do you need to make a row of 10 triangles? 100 triangles? *Any* number of triangles?

Imagine a square

Surround it completely with squares the same size; how many do you need?
How many squares are there altogether?

Imagine two squares joined together, edge to edge

Surround them completely with squares. How many do you need?

How many squares are there altogether?

Start with three squares in a row. Do the same thing . . .

Start with 100 squares in a row . . .

How many squares altogether?

Imagine a 2 by 3 rectangle

It is to be surrounded by a frame made of 1 by 1 squares. How many squares are there?

What about a 3 by 5 rectangle?

An 8 by 10 rectangle? *Any* size rectangle?

What about a 2 by 2 square? A 5 by 5 square? *Any* size square?

Imagine a 4 by 5 rectangle

It is to have squares all the way round the edge on the *inside*.

How many squares are there?

What about a 5 by 7 rectangle?

A 10 by 11 rectangle? *Any* size rectangle? *Any* size square?

Imagine a 3 by 3 square

How many little squares are there?

With a 3 by 4 rectangle, how many little squares?

Any size square or rectangle?

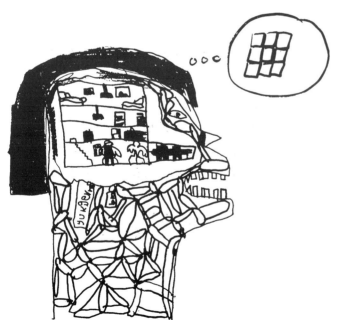

Imagine a 2 by 2 by 2 cube

How many little cubes are there?

What about a 3 by 3 by 3 cube? Other cubes?

Imagine a 2 by 2 by 3 cuboid. How many little cubes?

What about other cuboids?

Imagine a 10 by 10 by 10 box

It is to be filled with 1-unit cubes. How many are there on the bottom layer?

How many in 2 layers? 3 layers? 10 layers?

What about other size boxes?

Imagine a 3 by 3 by 3 cube

Its surface is to be covered with unit squares. How many squares are there?

What about a 4 by 4 by 4 cube? A 10 by 10 by 10 cube? *Any* size cube?

What about a 2 by 3 by 4 cuboid? *Any* size cuboid?

Imagine an equilateral triangle whose edge is 2

How many small triangles with an edge of 1 will fit inside it?

What about a triangle with edge 3? 4? More?

Imagine a 4 by 5 array of squares made from matches

How many matches are there?

How many intersections?

How many intersections where 4 matches meet? Where 3 matches meet? Where 2 matches meet?

What about *any* array of squares?

INFINITY

"What is the highest number?" is a question asked at some stage by many children, and stories about it are legion. It is facile to reply along the lines of: "There is no highest number, because I can add 1 to any number you give me". What is more interesting is to find out what ideas the children themselves have.

One thing about infinity is that, as has already been said, it is purely a mental idea; it is not something anyone can directly experience.

But one of the attractive things about infinity is that, in one sense, at primary school level it is not important; the National Curriculum does not mention it in connection with number; it is therefore not on the syllabus. And only much later on, and then only for some children, will it be necessary in mathematics. So in the meantime we can enjoy sharing children's ideas about it, and relish some of the paradoxes it involves!

One of the paradoxes is the number of whole numbers in relation to the number of even numbers. It seems reasonable, since every alternate whole number is even, to say that half the whole numbers are even, so there are more whole numbers. On the other hand, for every whole number I can double it to obtain a corresponding even number, and for every even number I can find a corresponding whole number by halving, so I can *match* the two sets; therefore the number of each is the same!

$$1 \rightarrow 2$$
$$2 \rightarrow 4$$
$$3 \rightarrow 6$$
$$4 \rightarrow 8$$
$$5 \rightarrow 10$$
$$6 \rightarrow 12$$
$$7 \rightarrow 14$$
$$8 \rightarrow 16$$
$$9 \rightarrow 18$$

. . . and so on for ever! (This is a characteristic of an infinite set. One definition of it is that it can be put into one-to-one correspondence with a subset of itself.)

Nine-year-olds, when asked for questions to which the answer is 10, have supplied sequences like

$20 - 10, 30 - 20, 40 - 30, 50 - 40, \ldots$

or

$20 \div 2, 30 \div 3, 40 \div 4, 50 \div 5, \ldots$

and said, "It goes on forever!"

Recurring decimals provide about the only explicit instance of infinity that primary children are likely to come across in the normal course of their mathematics. But even that is less likely than it used to be, since converting fractions to decimals is not required in the National Curriculum until level 6, and even then the examples given are only terminating ones. In any case the sensible use of a calculator for such conversions obviates the need for more than 7 decimal places.

So it can be interesting for children to consider an algorithm for, say, converting ⅓ to a decimal, because they will then discover the infinity of places involved. An algorithm can be developed using base-ten blocks, beginning with the block as a unit, and exchanging it for 10 flats.

Dividing by 3 gives 3 flats, with 1 left over to be exchanged for 10 longs, and so on. We get down to a unit left over, but by then we can see (mentally) that the unit can be treated like a new block, and the process can continue — for ever! Older children can convert this into a paper-and-pencil algorithm.

The Madison Project in the United States once produced a film of David Page working with some children on the question, "What is the highest number that is less than 3?" It obviously requires a knowledge of fractions or decimals to make it an interesting question!

"Infinity is more than you can imagine"

What is the highest number?

Think of some subtraction questions where the answer is 10

Think of some division questions where the answer is 10.

How many numbers are there between 1 and 10?

If you include fractions?

How many axes of symmetry has a circle?

How many points are there on a line?

How many odd numbers are there?

How many even numbers?
Are there more odd or more even?
Are there more whole numbers than even numbers?

Start with 1 and keep halving it

Where do you end up?

Divide 1 by 3, giving the answer as a decimal

Investigate other fractions as decimals.

What is the highest number which is less than 3?

GAMES

All these games can best be played with a class or group sitting in a circle, or in some other arrangement where the order of turns is clear. Or they can be played by a pair of pupils taking turns. Children can decide various rules for dealing with errors.

The games can be a vehicle for sharpening up children's mental agility, as well as for introducing some further ideas about numbers. But since the very idea of a game is to promote enjoyment, among other things, care should be taken that the level matches that of the children.

Count, starting at 1

Count in twos.
In threes.
In other ways.

Count in threes

But only the *units* digit is to be said.
For example:
3, 6, 9, 2, 5, 8, 1, 3, . . .
Try it with counting in sevens.
What happens with other rules?

Count as follows:

1;   1, 2;   1, 2, 3;   1, 2, 3, 4;   . . .

Count thus:

1, 2, 3, 4, 5;   2, 3, 4, 5, 6;
3, 4, 5, 6, 7;   . . .

Count in any of the ways suggested in the section on *Counting* on p.16

Count from 1

But every person who has a number in the 3 times table says "bing" instead:

1, 2, bing, 4, 5, bing, . . .

Now we start again, but every number in the 4 times table is replaced by "bong":

1, 2, 3, bong, 5, 6, 7, bong, 9, 10, 11, . . .

Try it with other tables.

Start again, but use two tables at the same time, for example 3s and 4s.

1, 2, bing, bong, 5, bing, 7, bong, bing, 10, 11, bing-bong, 14, . . .

Try it with other pairs of tables.

Try with *three* tables at once!

Start with 56

The next person has to give a divisor of the number, other than 1 or the number itself. If the number is prime, the next person adds 7.

For example:

56, 7, 14, 2, 9, 3, 10, 5, 12, . . .
or
56, 8, 4, 2, 9, . . .

Start with other numbers.

Make up similar rules.

Start with any number

The next person halves it if it is even, otherwise they multiply it by 3 and add 1.

For example:

26, 13, 40, 20, 10, 5, 16, 8, 4, 2, 1, . . .

(It soon becomes clear that after 1 you get 4, 2, 1, 4, 2, 1, . . . , so after 1 children can agree to begin with a new number.)

Do the same as before

But odd numbers are multiplied by 3 and then 1 is *subtracted*.

For example:

26, 13, 38, 19, 56, 28, 14, 7, 20, 10, 5, 14, 7, 20, . . .

You will have to decide where to stop these 'loops'. Maybe the first person to spot a loop shouts "loop!", and starts the next round.

Start with any number

(between 10 and 20 to begin with). The next person divides by 3 if the result is a whole number, otherwise must add *or* subtract 1 in order to make it a multiple of 3.

For example:

17, 18, 6, 2, 3, 1

When 1 is reached the next person starts with a new number.

(However, it is possible to discuss what happens if you continue after 1, since subtracting 1 gives 0, which *is* a multiple of 3.)

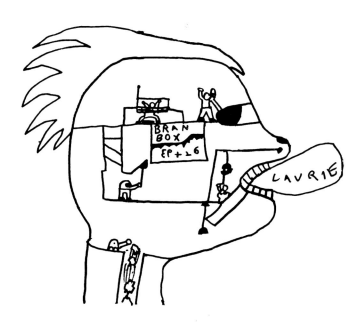

The mathematical brain at work

The activities in this book were trialled in the following Islington schools:

Penton Primary

Prior Weston Primary

St Mary's Primary

William Tyndale Primary

St Joan of Arc Primary